Letting Off Steam

Letting Off Steam

The Story of Geothermal Energy

A Carolrhoda Earth Watch Book

by Linda Jacobs

Carolrhoda Books, Inc./Minneapolis

Photo on page 2: Steam and other gases escape from a mud pot that lies deep in this hole in the Waiotapu Thermal Reserve near Rotorua, New Zealand.

Thanks to Stephen W. Robertson, Hydrogeologist, for his assistance with this book.

This edition of this book is available in two bindings:
Library binding by Carolrhoda Books, Inc.
Soft cover by First Avenue Editions
241 First Avenue North
Minneapolis, Minnesota 55401

LIBRARY OF CONGRESS CATALOGING-IN-PUBLICATION DATA

Jacobs, Linda, 1943-
    Letting off steam.

    "A Carolrhoda earth watch book."
    Includes index.
    Summary: Examines how geothermal energy, the force underlying hot springs, geysers, and mudpots, is being used as an alternative energy source in various parts of the world.
    1. Geothermal engineering—Juvenile literature.
2. Steam—Juvenile literature. [1. Geothermal engineering. 2. Steam] I. Title.
TJ280.7.J23 1989   621.44   88-6147
ISBN 0-87614-300-1 (lib. bdg.)
ISBN 0-87614-510-1 (pbk.)

Manufactured in the United States of America

2 3 4 5 6 7 8 9 10 99 98 97 96 95 94 93 92 91 90

Old Faithful's eruptions are watched and photographed by visitors from all over the world.

OLD FAITHFUL
GEYSER

In Wyoming's Yellowstone National Park, a crowd of excited tourists waits for Old Faithful to come to life. The geyser that erupts every 45 to 90 minutes is one of the park's most popular attractions.

Old Faithful starts its show with a small bubbling of hot water and steam. Then, as the tourists aim their cameras, a powerful spurt of steam and hot water shoots more than 100 feet (30 m) into the sky. This spectacular demonstration of geothermal energy amazes everyone who sees it.

Earth as seen from the Apollo 17 spacecraft. Nearly the entire coastline of the African continent is in view, including the Arabian Peninsula at the northeastern edge. Madagascar is the large island off Africa's southeastern coast.

Since ancient times, people have tried to understand and use geothermal energy. This is the story of what people have learned and of the many ways we have put that knowledge to work.

The word *geothermal* is the beginning of that story. *Geo* means earth or land; *thermal* means heat. **Geothermal energy**, then, is energy from our planet's heat.

The source of this heat is deep inside the earth, at its **core**, or center. The inner core is thought to be solid and is made up mainly of iron and nickel. The outer core contains the same metals but in a **molten**, or melted, state. It is hard to imagine just how hot the earth's core is. The temperatures there measure thousands of degrees on the Fahrenheit scale. Scientists believe that most of this extreme heat is produced by **atoms** as they **decay**, or break apart. Atoms are particles so small that they can be seen under an ordinary microscope only if there are 10 billion or more of them together.

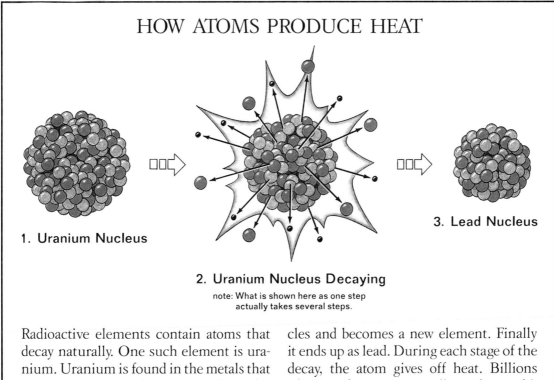

HOW ATOMS PRODUCE HEAT

1. Uranium Nucleus

2. Uranium Nucleus Decaying
note: What is shown here as one step actually takes several steps.

3. Lead Nucleus

Radioactive elements contain atoms that decay naturally. One such element is uranium. Uranium is found in the metals that make up the earth's core. When the **nucleus**, or center, of a single atom decays, it produces a small amount of heat. A uranium atom decays in a series of small steps. With each step, it loses some particles and becomes a new element. Finally it ends up as lead. During each stage of the decay, the atom gives off heat. Billions of atoms decaying naturally in the earth's core produce the extreme heat found in the core. Atoms can also be forced to break apart. In a nuclear power plant, atoms are split in order to produce nuclear energy.

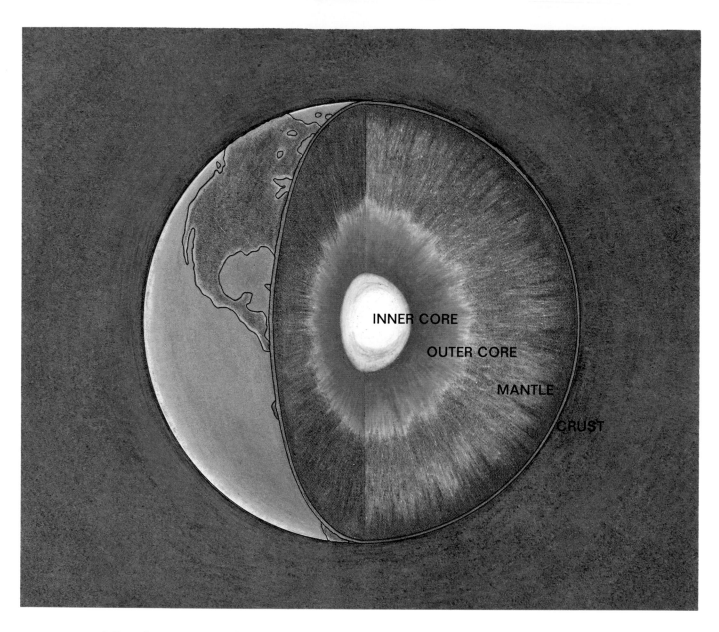

INNER CORE

OUTER CORE

MANTLE

CRUST

The hot core of solid and molten metals is surrounded by the **mantle**, solid rock that is also hot. The mantle is covered by a thin layer called the **crust**, which is the part of the earth we live on. Most of the crust is covered by soil, sand, and oceans. Although the earth is hottest at its center, the heat spreads outward through the mantle and the crust.

A miner guides his load through the gloom of a tunnel in the Homestake Mine, a gold mine deep in the earth's crust.

Geologists—scientists who study the earth—have learned that the temperature of the earth's crust increases with depth. If you were to go into a mine dug deep into the earth's crust, you would notice that the farther down you go, the hotter the temperature gets. This relationship of temperature to depth is called the **geothermal gradient**. The geothermal gradient is not the same over all the earth; some areas have high temperatures closer to the surface than other areas. There are different reasons for this temperature variation. In volcanic areas, the earth's crust is hotter because it, or the mantle below it, contains molten rock called **magma**.

The red-hot glow of lava being released from Kilauea, a volcano on the island of Hawaii, is evidence of the high temperatures deep within the earth.

In some volcanic areas, energy from the earth's heat is released in an erupting volcano as magma reaches the surface of the earth. Magma that comes to the surface of the earth is called **lava**. Lava can be as hot as 2,300° F (1,270° C).

A volcanic eruption is a form of geothermal energy, but it is much too powerful to control. A form of geothermal energy we can control is hydrothermal energy. *Hydro* means water. **Hydro-thermal activity** refers to hot water coming to the earth's surface from underground. It is both spectacular to see and useful in our daily lives.

WHAT IS STEAM?

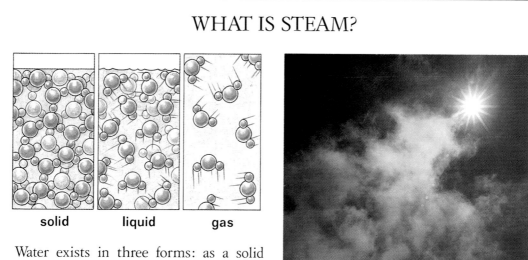

solid   liquid   gas

Water exists in three forms: as a solid (ice); as a liquid (flowing water); and as a gas (steam). The **molecules**, or individual units, that make up water each have two hydrogen atoms and one oxygen atom. No matter what form water takes, its molecules are always the same. Changes occur in the movement of the molecules, the amount of space between the molecules, and the forces binding the molecules. The molecules of a solid are locked together in an almost motionless formation. The bonds between molecules of a liquid are looser, and these molecules slide past each other in constant motion. The molecules of steam move so violently that they break the bonds between them. They are widely spaced and bounce off each other in a random manner.

True steam is an invisible gas. We cannot see the individual molecules that escape from the liquid water into the air. We say we see steam, but what we really see in the air are tiny drops of water that form as true steam cools.

Steam rises from hot springs at northern California's Bumpass Hell geothermal area in Lassen Volcanic National Park.

Wherever pockets of magma exist in the mantle and crust of the earth, the surrounding rock is heated. Such areas are likely places to find hydrothermal activity such as fumaroles, hot springs, mud pots, and, less frequently, geysers. These fascinating spouts of steam and bubbling cauldrons of water and mud result when underground heat, water, and pressure find openings to the earth's surface.

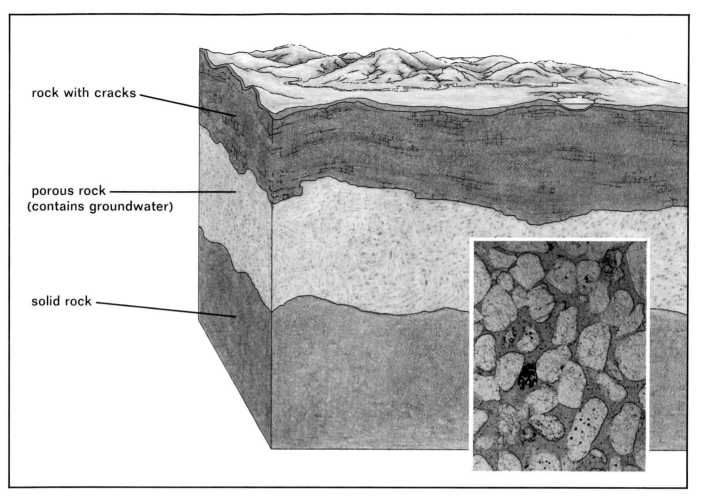

rock with cracks

porous rock
(contains groundwater)

solid rock

The inset photo shows a section of sandstone, a porous rock, photographed at 46 times its actual size. A bluish green stain that has been applied to the sandstone fills the pore space of the rock.

The hot rocks heat the **groundwater**, the water that lies underground. Rain and melted snow drain through the soil and down through cracks in rocks. The water is trapped in a layer of **porous rock** above a layer of solid rock. Porous rock is filled with small holes and holds the groundwater somewhat like a sponge. Groundwater is stored under pressure that comes from the weight of the rocks around it, the weight of the water itself, and the atmospheric pressure above it.

To be released from this pressure, the water must find a channel, or opening, to the surface. Such channels exist in underground rock that has many cracks. These cracks often appear in the underground rock of areas that have or used to have volcanoes or that have many earthquakes.

Since no one can see exactly what happens underneath a hot spring, a mud pot, a fumarole, or a geyser, scientists have formed **theories**, or explanations based on known facts, to explain their causes.

Groundwater heated under pressure by the rocks around it is extremely hot. Hot water is lighter than cool water, so the hot water rises toward the surface. According to the theories of most scientists, the water in hot springs, mud pots, and fumaroles rises freely through channels. These scientists believe that the amount of underground water determines the form the hot water will take as it rises to the earth's surface. If there is a lot of groundwater, the hot water will bubble to the surface as a clear **hot spring**. When there is not enough groundwater to clear away the mud that forms as the hot water dissolves rock, **mud pots** may form.

Above: *Clear water bubbles to the surface of a hot spring in Yellowstone National Park.*

Below: *Hot water and gases rise to the surface, dissolving rock and forming a mud pot in the Waiotapu Thermal Reserve near Rotorua, New Zealand.*

FLASHING TO STEAM

A flash is a brief exposure to a strong changing force, such as extreme heat or cold. Water is said to flash to steam when it changes from liquid water to vapor in an instant. One example is the "disappearance" of a drop of water when it hits a very hot piece of metal, such as a drop of water evaporating the instant it touches the surface of a hot frying pan.

Left: *Steam and other gases are released from this fumarole in Yellowstone National Park.*

Most scientists believe that **fumaroles** result from very little water in the underground system. What comes out of a fumarole's vent is a mixture of gases—steam, carbon dioxide, and some hydrogen sulfide. These gases are forced out when small amounts of cold water that seep from above ground hit hot rock and **flash**, or change, into steam. When this happens, the steam takes up more than 1,600 times the space that the water took up, which forces the gases up and out through the vent. On the surface, we see a continuous release of steam as it hits the air and cools.

STEAM EXPANDS

When the tightly packed molecules of a liquid are heated, they increase their movement until they escape into the greater space of the atmosphere. When water boils, it turns into steam that occupies about 1,600 times the space the water took up. If the steam is heated still further, it expands to take up even more space. The force of this expansion is behind the power seen in geysers and used in steam engines and turbine engines.

A much rarer form of geothermal energy is a geyser. Fumaroles, hot springs, and mud pots are found in geothermal regions all over the world. Geysers are found only in Iceland, New Zealand, Indonesia, and the United States. Yellowstone National Park in Wyoming is the only place geysers are found in the United States. **Geysers** are hot springs that seem to turn on and off. Unlike hot springs that gush forth hot water continuously, geysers seem dry except when they erupt and send jets of hot water and steam skyward.

There are many theories that have been formed to explain the workings of a geyser. The most generally accepted theory explains

Hot water and steam shoot skyward during an eruption of the Prince of Wales Feathers Geyser near Rotorua, New Zealand.

that a geyser's underground channels are different from the channels of a fumarole, a mud pot, or a hot spring. Underneath a geyser, water collects in a system of narrow, twisting channels and small pockets. Most of this water is hot water that circulates close to the magma then rises to fill the passageways. Some of the water is cool water from the surface that collects in the channels and goes no farther. Because the trapped water is under great pressure, it heats up far beyond its normal boiling point without boiling. Such water is called **superheated**.

SUPERHEATED WATER

When a liquid is heated, the molecules move faster and faster. The molecules, normally bound to each other, move so much that they break their bonds and change into **vapor**, or steam. **Vapor pressure** is the pressure steam exerts above a liquid. **Atmospheric pressure** is the weight of the molecules that make up our air or atmosphere. As a liquid gets hotter and molecules move faster, vapor pressure increases. When the vapor pressure equals the atmospheric pressure, the liquid boils.

High on a mountain, the atmospheric pressure is low because there are not as many molecules packed into a square inch of air as there are at lower land levels. At sea level, the atmospheric pressure is greater than it is up on a mountain. Below sea level, the number of molecules per square inch of air is even greater. When water is heated at sea level, it boils at 212° F (100° C). If the atmospheric pressure is greater than it is at sea level, water takes longer to boil than it would at sea level. It takes longer for the vapor pressure to match the higher atmospheric pressure. Water heated under these conditions gets hotter than 212° F (100° C) before boiling. This water is called superheated water.

The steam bubbles created by the superheated water rise toward the surface and clog the narrow channels. Eventually the steam bubbles force some of the water in the system to the surface. When this happens, the pressure is lessened on all the water in the system. The water then flashes to steam that takes up thousands of times the space that was filled by the water. The only direction the steam can go is up.

This geyser bubble is about to erupt. In the diagrams showing the underground workings of a geyser, the groundwater circulates and collects in the underground pockets (1). The steam bubbles force the water upward (2). The release of pressure allows the water to flash into steam, carrying more water with it (3).

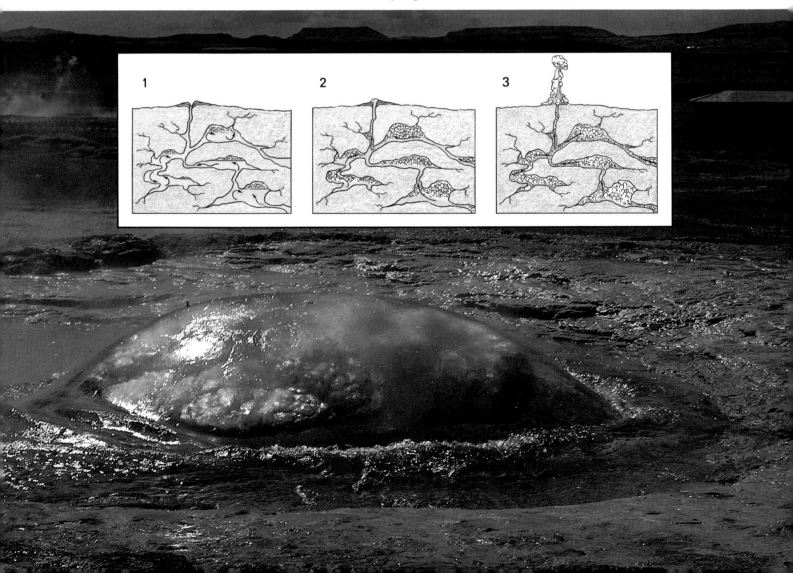

The word geyser *is derived from* Geysir, *which is the name of this geyser, located 47 miles (75 km) northeast of Reykjavik, Iceland.*

As steam shoots out of the narrow channel above the pockets of water, some of the water from the pockets is carried with it. This reduces the pressure, allowing even more water to turn to steam and erupt, carrying more water with it. This goes on until all water from the pockets has been forced to the surface. After the eruption, the groundwater pockets begin to fill with cool water that seeps down from the surface and with hot water that rises from the porous rocks below the surface. The process starts all over again.

The eruptions of Old Faithful, one of several geysers in Yellowstone National Park, are predictable. In 1938, a park ranger discovered that the amount of time between eruptions is based on the length of each eruption. The longer a geyser spurts steam and water, the more water it forces out. The underground channels then take longer to refill with water.

Although Old Faithful does not erupt every half hour or even every hour, it is "faithful" to a predictable schedule.

Normally geysers, such as these in this geyser basin in Yellowstone National Park, have interconnected underground systems.

Most geysers do not keep to this sort of schedule. Because the superheated water dissolves the rock, the channels are constantly changing. The amount of time it takes to fill the channels cannot be predicted. Geysers are also sometimes linked to each other's underground systems. This means the amount of water in each system at any given time is unpredictable. Old Faithful, with an independent underground system in rock that is not easily dissolved, keeps to a schedule that has varied only slightly since it was discovered and named over 100 years ago. The amount of time between eruptions ranges from 45 minutes following a 1- to 2-minute eruption to 90 minutes after a 5- to 6-minute eruption.

The **plume**—the stream of water and steam—from a geyser can shoot more than 200 feet (61 m) into the air. The height of the plume depends upon the amount of water in the underground pocket.

The water temperature in this hot spring in the Waimangu Thermal Valley near Rotorua, New Zealand, encourages the growth of green algae. The reds, yellows, and browns are caused by the chemical reaction of the iron with the rocks and water.

Hot water bubbling, hissing, and erupting to the surface of the earth is an amazing sight. Hot springs display many different colors. The color is caused by algae, bacteria, and minerals. If the water in a hot spring is very hot, about 170° F (77° C), the algae are usually yellow. In cooler water, the same algae might look orange, red, brown, or green. Yellow or pink strands of bacteria can sometimes be seen in hot springs. Sulfur, iron, and other minerals also add color to the pools of hot water.

The vent of Giant Geyser in Yellowstone National Park is surrounded by water that contains minerals and algae.

Cones made of silica have built up around the geyser vents of the Lion's Group, a gathering of geysers in Yellowstone National Park. Silica is a mineral that is contained in rhyolite, the volcanic rock that underlies most of the Yellowstone area.

Groundwater contains minerals that come from the rock through which the water passes. Hot groundwater dissolves more rock than cool water, so it has a very high mineral content.

Mammoth Hot Springs in Yellowstone National Park is a terrace of travertine, a kind of limestone. It is formed as the hot waters of the spring bring up large quantities of dissolved limestone from the limestone beds that underlie this area.

The high mineral content of geothermal water can be seen not only in the colors that the dissolved minerals add to the pools of hot water, but also in the solid mineral deposits that the water leaves behind. The kind of rock that the water is stored in and passes through on the way to the surface determines how the mineral deposit will look.

The barren appearance of this geyser basin in Yellowstone National Park is due to the scalding waters and high concentrations of minerals and gases that are harmful to plants.

Areas with a lot of hydrothermal activity do not all look the same. Some areas, such as the geyser basins in Yellowstone National Park, have almost no plant life. Most plants cannot live where they are always threatened by scalding waters and high concentrations of minerals. Some of the smelly gases that come from underneath the ground, such as those containing sulfur, are also harmful to plants. The Yellowstone geyser basins smell like rotten eggs—the smell of sulfur compounds.

Fumaroles surrounded by lush plant growth steam from behind rocks in the Waimangu Thermal Valley in Roturua, New Zealand.

In some areas with hydrothermal activity, fumaroles and hot springs are surrounded by plants. It could be that the water released to the surface in these areas is cooler or that it contains fewer dissolved minerals than the water that surfaces in the areas where no plants will grow.

Hydrothermal activity is not only spectacular and beautiful, it is also a very useful natural resource. For centuries, people have been using hot water from the ground as a source of heat.

The oldest written description of a geothermal field comes from the ancient Romans. About 2,000 years ago, a writer described a huge field crowded with fumaroles and smelling of sulfur. This area in Italy is now called Larderello. The Romans constructed hot spring baths near this area. People came to soak away their aches and pains in the hot mineral waters. Today, many resort areas depend on the soothing comfort of hot mineral waters to attract visitors.

Vacationers enjoy a soak in this enormous hot springs pool at Glennwood Springs, Colorado.

A hot spring bubbles out of a rocky hillside near Steamboat Springs, Colorado.

The native people of New Zealand, the Maoris, used the hot waters of the many geothermal regions in their country to cook food. In New Zealand today, people continue to use the heat of geothermal water. They now cap springs and pipe the water into constructed pools and even into heating systems. In this way they can use water from hot springs that empty onto level or rocky ground instead of into a natural pool. Untapped, this water would drain back into the earth or quickly cool to the temperature of surface waters.

Near New Zealand's Wairaeki field, water from a hot spring is piped around a hotel swimming pool. The pipes that carry the thermal water touch the outer walls of the swimming pool. The pool itself is filled with cold water, but the pool water is kept warm by geothermal water that nobody ever sees.

An iceberg floats in the meltwater of a glacier, an indication of the arctic climate in Iceland.

Like the Romans and the Maoris, the first people of Iceland used the hot water from the ground that is so plentiful on their island. They washed clothes in geothermal water and used the water as a heat source for cooking. Then, in the thirteenth century, a poet and historian named Snorri Sturlvson thought of a new way to use the hot water. Snorri lived near a hot spring, and he cut channels into rock so that hot water from the spring flowed around his house. He put in pipes to circulate it through his walls. Snorri is credited with being the first person to heat a house with hot water from the ground.

Heat is especially important to Iceland, a tiny island country near the Arctic Circle. A "warm" day there is barely over 60° F (16° C). A cold day is many degrees below freezing. So it is surprising that Snorri's ideas were not put into general use until over 700 years later.

Plants that do not normally grow in Iceland's cold climate can grow in this geothermally heated greenhouse just outside of Reykjavik.

Today, Iceland's capital city of Reykjavik (rayk-yah-vick) gets 90 percent of its heat for homes and public buildings by directing hot water from the earth into a system of pipes under the streets. Natural hot water is also used in swimming pools, washing machines, and showers. Large greenhouses heated by the earth's hot water make it possible for Icelanders to grow fruits and vegetables that would otherwise never survive in the bare, cold countryside.

The port town of Reykjavik, Iceland, lies close enough to its hot water resources to use them directly.

To be a practical heat source, hot springs must supply a lot of water, and that water must be hot enough to use. Some springs are just trickles, and some bring up so little water that they don't flow as liquid at all. Iceland is an actively volcanic island and has some of the most spectacular displays of hot water features in the world. Icelanders are able to use the earth's hot water directly because there is so much of it all over the island. Using natural hot water for its direct heating properties is not possible unless the source of the hot water is relatively close to where it will be used. Water loses heat when it is piped very long distances.

This engraving, entitled The Gates of Hell, *was done by William Bell Elliot upon his 1847 discovery of a geothermal field in California. Elliot named the field The Geysers, mistakenly identifying the fumaroles that dotted the area as geysers.*

Many people who found ways to use hot springs were afraid of the power of fumaroles and geysers. For some, the hot, hissing, steaming environment where geysers and fumaroles are found resembled their visions of Hell. For a while, fear and lack of modern technology stopped people from finding new uses for the hot water energy below the surface of the earth. Today, geothermal water is still used as a direct source of heat, but it also has a new use—to produce electricity.

Electrical energy causes lightning and gives us occasional shocks. When it is controlled, electricity powers our toasters and light bulbs. The electricity we use every day comes from power plants. Power plants don't actually create electrical energy; they transform other types of energy into electrical energy. A power plant does this by changing **kinetic energy**, or the energy of motion, into mechanical energy in order to **generate**, or produce, an electrical current. An electrical current is a controlled form of electricity that can be used to power machines and appliances.

This power plant burns coal to heat water and make steam that will be used to run its turbine engines.

To run the machines that generate an electrical current, a power plant needs a source of kinetic energy—something to move its **turbine engines**. Turbine engines often look like huge pinwheels, with curved blades arranged around a central shaft. When a force hits the blades, it causes them to spin, turning the shaft. The shaft then rotates parts in the **generators**, which are the machines that actually produce the electrical current.

A worker inspects the blades of an enormous turbine engine.

Some power plants, called **hydroelectric plants**, use rushing water from a dammed-up river or lake to provide the necessary force to power their turbine engines. Many more power plants, though, use the release of steam under pressure as their source of kinetic energy. In many power plants, the steam is produced by heating water. The fuels used to heat the water are most often oil and coal, but natural gas and nuclear energy are also used. A geothermal power plant does not use any of these fuels to produce steam. The earth has already heated water that, once released, will flash to steam.

Builders of a geothermal power plant must drill many deep wells to bring up the hot water or the steam. Different geothermal fields have different temperatures. The temperature of the underground water determines what kind of equipment is needed to operate the geothermal power plant.

A well at The Geysers controls the release of steam from the ground.

The best fields for producing electricity are so hot that they produce mostly steam with very little liquid water. These fields, called **dry steam fields**, require the least amount of equipment, since the steam from the well can be used directly to run the turbine engine. Cooler fields have a nearly equal mixture of steam and water. These **wet steam fields** require equipment to separate the steam from the water before the steam can be used. **Hot water fields** use a heat exchange system to generate electricity. Since the water is not hot enough to produce an amount of steam that would effectively operate the turbine engines, the water is used to heat a fluid that turns into vapor at a lower temperature than water. Fields that use this heat exchange system usually have water that is between 320°F and 400°F (161°C and 206°C). Fields with water that is cooler than 320°F (161°C) are not usually developed to produce electricity.

Although the three types of systems are different in many ways, they all use wells to bring up the water or steam, and pipes to carry it to the nearby power plant. Tapping into a hot water source and piping it into a plant sounds simple, but finding and using hydrothermal resources involve much research and equipment. Like all natural resources, hydrothermal resources must be used carefully. Developers of geothermal power plants must remember that the conditions of heat, water, and pressure that make an area right for producing steam have to be kept in balance.

Insulated pipes extend from a producing well at The Geysers. The Geysers is a dry steam field, so the pipes carry mostly steam back to the turbine engines of the power plant.

Steam rises from the cooling towers at The Geysers. After the steam has been used to drive the turbine engines, it is run past cool water pipes so that it will condense into liquid. The water that has absorbed the heat of the steam in order to turn it back into liquid is then piped to large towers to cool so that it can be pumped back into the ground.

People who live near a geothermal power plant have a convenient, clean, reasonably inexpensive source of electricity. Even those who live too far away from a geothermal power plant to use its electricity benefit from it. Because the plant doesn't use oil, natural gas, or nuclear material to generate electricity, these fuels are available for use elsewhere.

When the earth lets off steam, mud pots bubble, geysers erupt, and hot springs flow. Since ancient times, people have used the earth's heat to soak their bodies, to warm their homes, and to cook their food. Now we're using it to produce electricity. In the future, new ways of finding and using geothermal energy may be discovered. More people will then benefit from the power that makes flowers grow in frozen Iceland and turns on lights all over the world.

It is hard to see through the steam that rises from the surface of this large, boiling mud pot in Yellowstone National Park.

Hot springs in Yellowstone National Park deposit layers of limestone, forming an ever-growing terrace as the water continuously rises to the top.

Photo Credits

Photographs courtesy of: pp. 2, 18, 19, Milton T. Pasley; pp. 5, 15 (top), 26, David Falconer; p. 6, NASA; p. 9, Homestake Mining Company; p. 10, Greg Vaughn/Hawaii Visitors Bureau; pp. 11, 25, 47, Larry Day; p. 12, C.W. May/BPS; p. 13, American Association of Petroleum Geologists; pp. 15 (bottom), 24, 29, 44-45, Robert E. Cramer; pp. 16, 23, Bill Johnson; pp. 17, 27, © Linda Robinson; pp. 20, 21, B.F. Molnia/Terraphotographics/BPS; pp. 22, 28, Kahnweiler/Johnson; pp. 30, 31, Paul A. Pavlik; pp. 32, 33, 34, Leonard Soroka; p. 36, Jerry Miller/Northern States Power; p. 37, Mitch Kezar; pp. 39, 42, Marla Murphy; p. 41, Gerald A. Corsi/Tom Stack & Associates; p. 43, Willma Willis Gore

Front and Back photographs courtesy of Alden Berry

Engraving on page 35 courtesy of Pacific Gas and Electric Company

GLOSSARY

atom: a tiny particle that is considered to be a source of potential energy

core: the center of the earth

crust: the thin layer of the earth on which we live

decay: to break apart and produce energy

dry steam field: a hot geothermal area that produces steam with very little liquid water

flash: a brief exposure to a strong changing force, such as when cold water hits a hot surface and flashes to steam

fumarole: a hole from which steam mixed with volcanic gases and vapors escapes when there is very little groundwater

generate: to produce

generator: a machine that changes mechanical energy to electrical energy

geologists: scientists who study the earth

geothermal energy: energy from the heat of the earth's interior

geothermal gradient: the relationship of temperature to depth in the earth

geyser: a spring that is at times dry and at times sends jets of hot water into the air

groundwater: surface water that lies in the part of the ground that is fully saturated

hot spring: a spring of water heated naturally within the earth

hot water field: a field where water is used to heat a fluid that turns into vapor at a lower temperature than water

hydroelectric plant: a power plant that produces electricity by waterpower

hydrothermal activity: hot water coming to the earth's surface from underground

kinetic energy: the energy of motion

lava: magma that comes to the surface of the earth

magma: molten rock

mantle: hot solid rock that surrounds the earth's core

molten: liquified by heat

mud pot: the form hot water takes when it rises to the surface in areas where there is not much groundwater

plume: the stream of water and steam from a geyser

porous rock: rock filled with tiny holes

superheated water: water that heats up beyond its normal boiling point without boiling

theory: an explanation based on facts and the way they relate to one another

turbine engine: a machine with curved blades arranged around a central shaft that is turned by moving fluid. This action changes kinetic energy into mechanical energy.

wet steam field: a geothermal field that produces a nearly equal mixture of water and steam

INDEX

ABOUT THE AUTHOR

Linda Jacobs has written more than 60 books for young people—both fiction and nonfiction. While working as a public information writer at the Naval Civil Engineering Laboratory in southern California, Ms. Jacobs wrote on various scientific topics. Her interest in geothermal energy developed when she moved to Lake County, site of The Geysers. Ms. Jacobs lives in Clearlake, California, with her husband, Richard Altman.